Seaweeds of Cape Cod Shores

A FIELD GUIDE

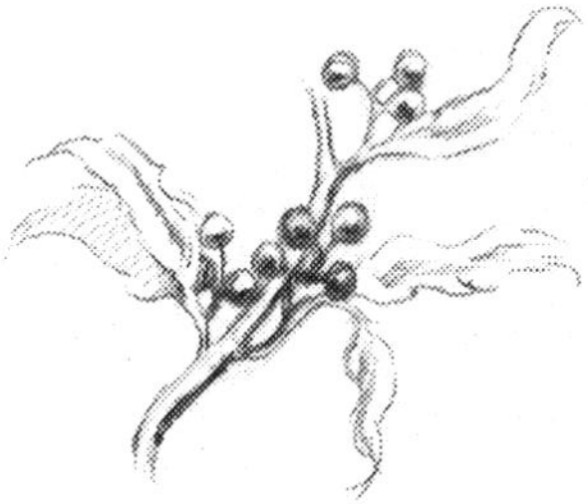

by Gilbert Newton

with illustrations
by Nancy Minnigerode

Gil Newton

Published by Gilbert Newton

Fourth Printing 2022

Printed in the USA
West Barnstable Press
ISBN 0-9816873-0-8

Cover photograph by Nancy Viall Shoemaker

Table of Contents

Introduction 1

Acknowledgements 3

The Nature of Seaweeds 4

Save the Seaweeds 5

The Green Algae

Chaetomorpha linum 6
Cladophora sp. 7
Codium fragile 8
Enteromorpha (Ulva) intestinalis 9
Ulva lactuca 10

The Brown Algae

Agarum cribosum 11
Ascophyllum nodosum 12
Chorda filum 13
Chordaria flagelliformis 14
Ectocarpus siliculosus 15
Fucus vesiculosus 16
Leathesia difformis 17
Laminaria saccharina 18
Petalonia fascia 19
Scytosiphon lomentaria 19
Sargassum filipendula 20

The Red Algae

Agardhiella tenera 21
Gracilaria sp. 21
Bonnemaisonia hamifera 22
Ceramium rubrum 23
Chondrus crispus 24
Corallina officinalis 25
Dasya pedicellata 26
Grinnellia americana 27
Polysiphonia sp. 28
Porphyra umbilicalis 29
Palmaria palmata 30

The Flowering Plants

Zostera marina 31

Recommended References 32

Biographies 34

Introduction

"Which one is this? It's very slimy!"

"I can't tell one from the other. They all look the same."

"This one is a pretty red, at least when it's underwater."

"The only one I recognize is the one where you pop the swollen tips."

"Can you eat this one?"

These are some of the comments and observations my botany and coastal ecology students have made over the years whenever we study the seaweeds. This group of marine organisms is often baffling and complicated to most people. Though there are several excellent books and field guides available that include seaweeds, they are usually not for the novice collector or beachcomber. Many use technical language that is difficult to understand without a background in botany, marine biology, or phycology, the scientific study of algae. Many books are so comprehensive that they include tropical species listed with those from temperate waters, thus creating even more confusion.

Even though I love any book about seaweeds, I decided to write this small field guide for students on Cape Cod to make it user – friendly and simple as possible. For many years I have been taking my high school and college students to the seashore to identify and study the diverse species of marine algae in these waters. My goal here is to provide a quick field reference for the beginning student of all ages who can use the text and drawings of Nancy Minnigerode to identify some of the most common marine algae species in this region. We show the illustrations as large as possible, but please note they are not drawn to scale.

It is always my hope that the students will develop an interest in seaweeds and pursue additional studies of them. With that in mind, I have included a list of other books on this topic. The reader can use these more detailed resources for a lifetime of discovery and enjoyment of this remarkable group of marine life.

Acknowledgements

I would like to thank the following for their gracious assistance in this publication:

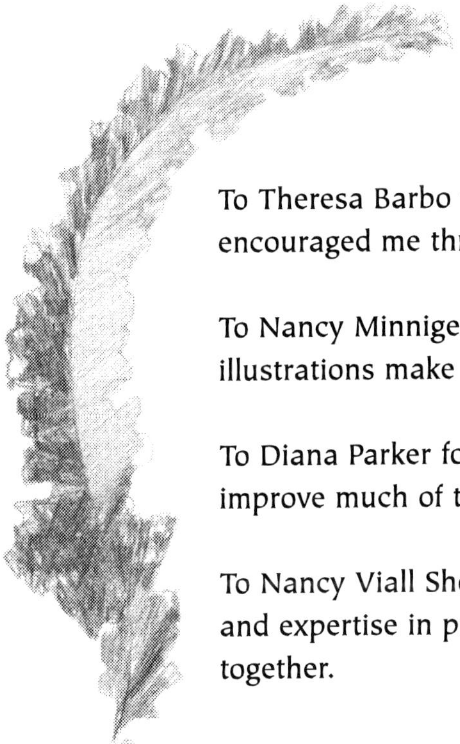

To Theresa Barbo who inspired and encouraged me through the entire process.

To Nancy Minnigerode whose magnificent illustrations make this book possible.

To Diana Parker for helping to clarify and improve much of the text.

To Nancy Viall Shoemaker for her advice and expertise in putting this publication together.

And finally to all of my students, past and present, for whom this book is dedicated.

The Nature of Seaweeds

The word seaweed refers to a group of marine organisms called macroalgae. Most people think of them as plants. Technically they belong to a different kingdom called the Protista. Algae lack the specialized tissues for the transport of food and water. Like the plants, they do photosynthesize, making their own food from sun energy. However, they don't have any true leaves, stems, roots, flowers, fruits, and seeds. They do have other structures which resemble these familiar plant parts, and even carry out similar functions. Thus, we describe an alga's leaf-like blade, its stem-like stipe, and its root-like holdfast.

Reproduction in algae can be by asexual methods, such as fragmentation (a new alga growing from a piece of the parent) or spore production, or sexual methods in which male and female gametes are released into the water for fertilization.

In this guide, the seaweeds are divided into three groups: the greens (*Chlorophyta*), the browns (*Phaeophyta*), and the reds (*Rhodophyta*). This classification is based primarily on the different pigments present in the cells of the algae. Botanists call these groups phyla or divisions.

Seaweeds are important both ecologically and economically. Many small animals, such as periwinkles, amphipods, and bryozoans use seaweeds as microhabitats where they graze, avoid the sun, and escape from predators. Seaweeds are photosynthetic, so convert sun energy into sugars and oxygen, and are the base of the marine food web. Eventually this energy is passed on to the next feeding levels, and supports the larger fish populations.

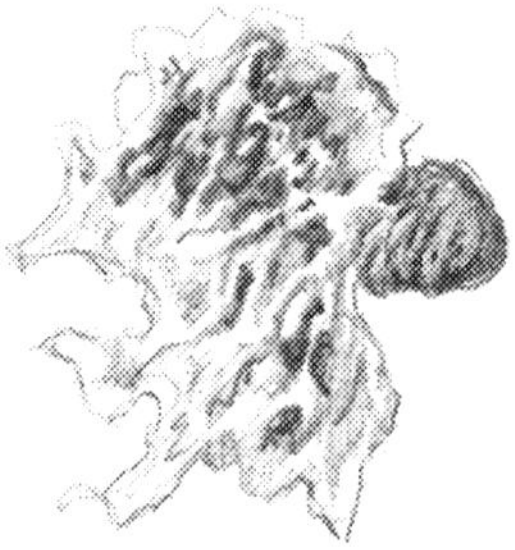

Seaweeds are widely used in industry and agriculture. Extracts from algae such as Irish moss and kelp are used as stabilizers in several food items. Some algae are also used as food for cattle and chickens. Many species contain vitamins, protein and other nutrients. Seaweeds have also been used in medicine, cosmetics, and as a source of fertilizer.

Save the Seaweeds

Many of the common seaweeds can be preserved for future reference and study. This method can also be used for various arts and crafts projects using seaweeds, and the materials needed are quite simple. The easiest ones to preserve are the stringy, flat, or filamentous forms. The large, thick ones require more complex methods for preservation.

ACTIVITY:

1) Visit the beach and look for thin, filamentous seaweeds that have washed up, or are present at the water's edge.
2) Make sure that the plants still retain their color. Algae will bleach out their pigments if exposed to the sun for a long time after being stranded on the beach.
3) Collect several of the plants in a bucket, and fill the bucket half-way with sea water.
4) Back in the classroom, or at home, obtain a tray or pan large enough for the plant to be spread out.
5) Fill the tray with sea water and float the plant in it.
6) Use a thick sheet of paper, such as official herbarium stock, oak tag, or even a large index card.
7) Place the paper in the water and under the plant.
8) Slowly raise the paper so that the plant is picked up.
9) Carefully drain the excess water from the paper.
10) Using a medicine dropper, squirt water around the seaweed to arrange it the way you want to see and to remove small pieces and debris that may remain.
11) Cover the plant with cheese cloth or wax paper.
12) Place between several sheets of newspapers, and place several large books on it. (Note: Small plant presses can also be purchased for this stage as well).
13) The newspapers should be changed the next day because they absorb the water.
14) In 2-3 days, carefully peel off the cheese cloth or wax paper, and the seaweed will remain stuck to the herbarium paper.
15) You should label your specimen with information that includes its name, habitat, and date of collection.

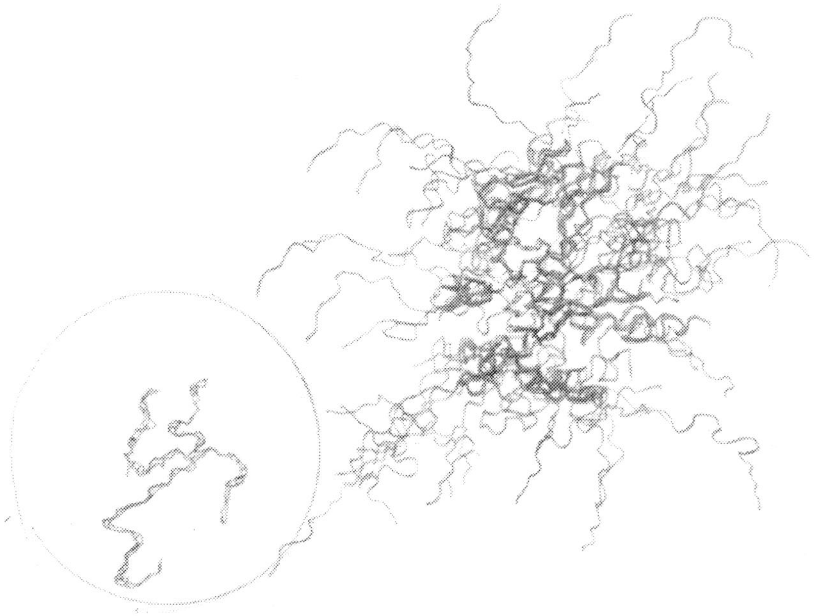

Chaetomorpha linum

You should handle this green alga to identify it for it feels and looks like tangled green threads. Coarse to the touch, this alga is often seen floating along the shore or in filamentous mats in tide pools. Its wiry texture and unbranched filaments will enable you to distinguish it from other green algae in the same habitat.

It can be found in warmer water, and I often collect it wrapped around other algae in shallow waters. I have seen it in large mats, particularly during the summer. Use of a hand lens may reveal rows of cells along the filaments.

Cladophora sp.

Cladophora is a green alga in which it is very difficult to identify the species with just a hand lens. It can also be easily confused with other green algae. The color and texture can vary, but look for a fine, light green, filamentous alga that is branched, and can be more than a foot long.

This alga is also an environmental indicator for nitrogen runoff, and can be a major nuisance in bays, estuaries, and quiet bodies of water. It forms large, tangled mats that collect and decompose along the shore.

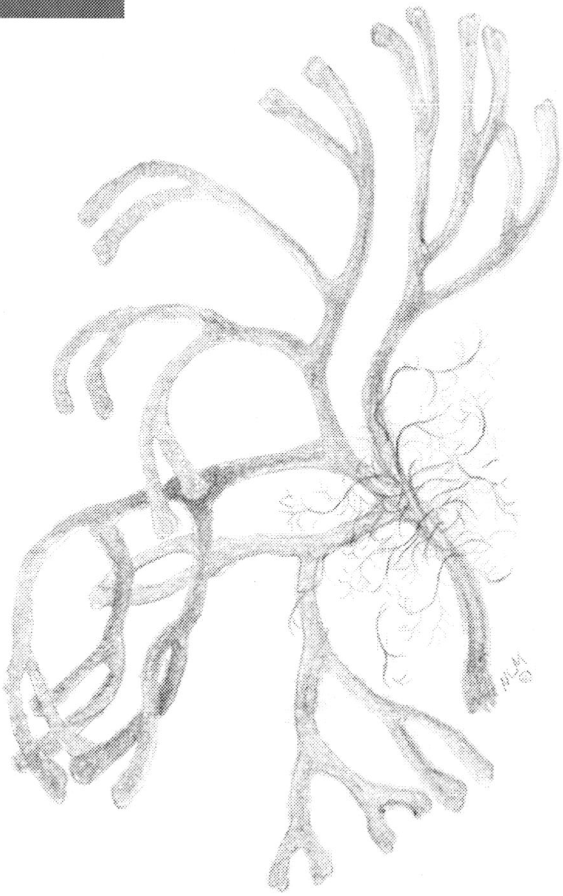

Codium fragile (Green fleece)

This introduced species is, unfortunately, very common along sandy beaches. This alga has a spongy texture, and its branches have puffy, swollen tips. It can grow up to three feet in length, but often forms large, heavy mats and clumps that frequently wash up with the tides, and can dominate the wrack line along the beach.

Codium prefers cold water, and can be a major pest in shellfish beds. It attaches to the shells of oysters, scallops, and slipper snails, weighing them down, and sometimes covering their filter-feeding siphons. *Codium* spreads easily by fragmenting into new clumps, and it is difficult to control.

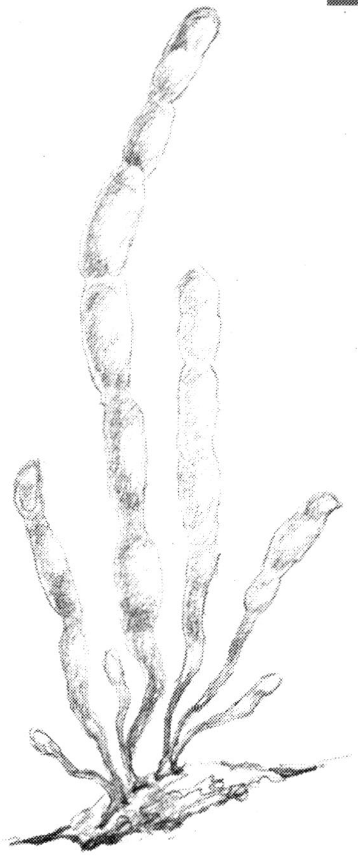

Enteromorpha* (*Ulva*) *intestinalis
(Hollow or tubular green weeds)

One of the easiest seaweeds to collect, identify, and press is the green alga *Enteromorpha*. There are several species that can be found in the upper intertidal zone, but one that is easily identified is *E. intestinalis* because of its shape. Looking like green human intestines, the blades are long, unbranched, and filled with a gas that gives it a swollen appearance.

Recently, the genus name has been changed to *Ulva*, and like the sea lettuce, this alga can tolerate changes in salinity (salt content) and higher levels of the nutrient nitrogen. At low tide, this is usually the most abundant species available and makes a very nice pressed specimen.

Ulva lactuca (sea lettuce)

One of the most common examples of green algae is *Ulva lactuca*, referred to as sea lettuce because it resembles salad lettuce. Consisting of large green sheets that often fold along the edges, this alga is found in the intertidal zone, and washes up on many beaches. This species is two cell layers thick, and can grow up to three feet in length.

Sea lettuce can grow in polluted bays and estuaries. It is often tolerant to changes in salinity, and is abundant where nitrogen runoff, acting as a fertilizer, is present. It is supposedly edible, but it doesn't appear appetizing because of its growth in these polluted waters, and the fact that there are often many small animals attached to it.

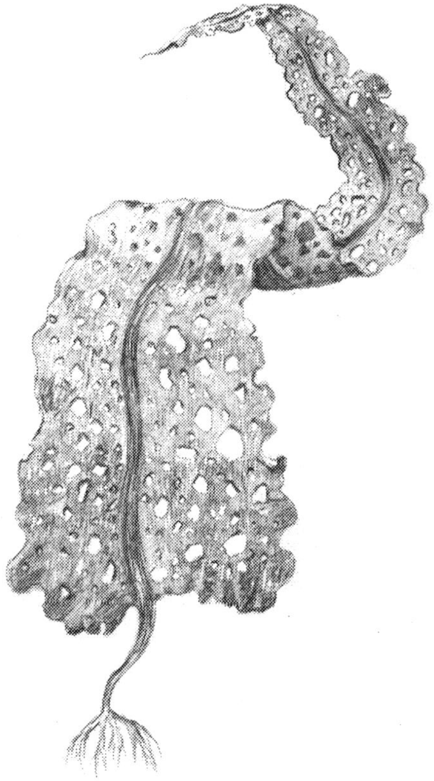

***Agarum cribosum* (sea colander)**

This large, conspicuous brown alga is found attached to rocks in cold, deeper water, but may be found in the tidal drift. It gets its common name from its unique appearance. The alga consists of a fan-shaped blade up to a foot wide that is riddled with holes. It looks like some herbivore has been munching on the alga, but its perforations are due to its growth habit and the effects of waves.

I have found this to be more common at a rocky shore, but I have collected samples near jetties and groins where the algae have been able to attach their strong holdfasts. This seaweed is also called the Swiss cheese alga.

Ascophyllum nodosum (Knotted wrack)

A familiar and robust brown alga is *Ascophyllum nodosum*, commonly called knotted wrack. This large rockweed has branches that can grow up to three feet long. The fronds are characterized by several large air bladders, but no distinct midrib.

This is a common alga growing in the intertidal zone, and is attached to rocks and jetties by its holdfast. It is the most conspicuous seaweed growing on the rocks in the Cape Cod Canal. There is a vegetative form that can be found in salt marshes. *Ascophyllum* has been used in clambakes, and as a fertilizer in gardens.

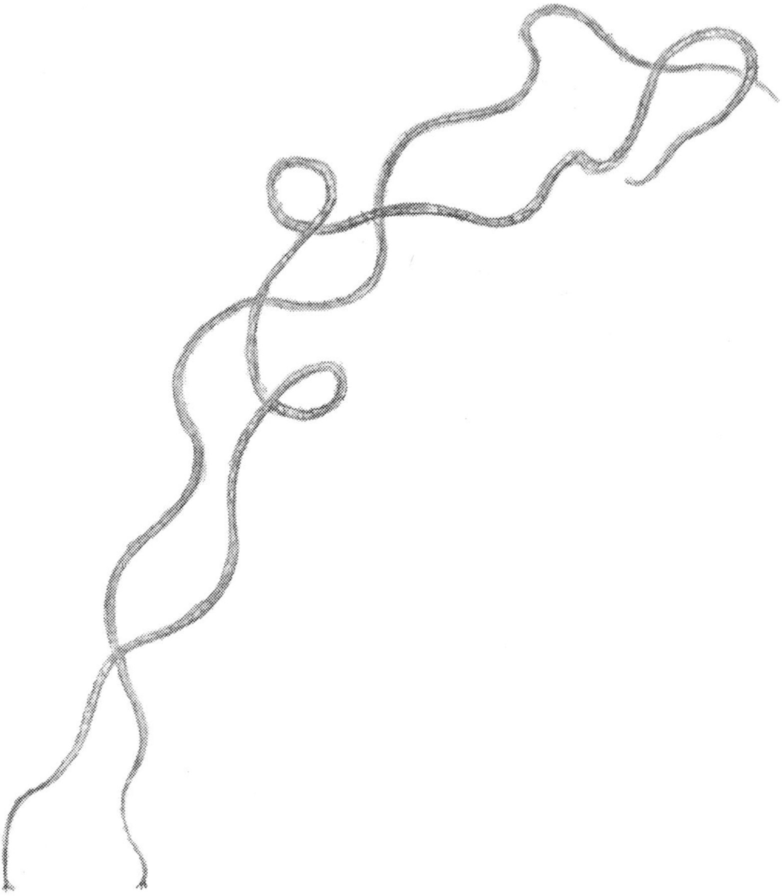

Chorda filum (sea lace)

Chorda is a long, dark brown, whip-shaped alga that is attached to rocks or shells by a small flat holdfast. This brown alga is in the kelp group, though it looks quite different from its relatives.

Chorda appears in the summer and grows in deeper water, often in clumps. It feels smooth when handled, and is unbranched. Some specimens may be greater than fifteen feet long.

Chordaria flagelliformis

This brown alga is found in deeper, subtidal water. The alga is a very dark brown to almost black color, and can grow up to two feet. It is sometimes deposited along the beach with the drift at high tide.

It can be identified by its smooth texture, but also by its branching. There is a main branch, and many short ones at the lower end. As you look towards the top of the plant, you will observe longer, unbranched sections.

This is one of the species that can be more easily located by visiting the end of a rock jetty where it is sometimes attached, along with several other brown algae species.

Ectocarpus siliculosus

One of the finest filamentous brown algae species is *Ectocarpus*. This very common alga is found in quiet bays, and is often attached to eelgrass or other algae. When removed from the water, it feels like a slimy, soft mass of filaments.

To see its unique reproductive structures requires the use of a microscope. But a good hand lens will reveal its many thin branches. This is one of the softest brown algae species, and it is sometimes confused with red algae because of its appearance and texture.

Fucus vesiculosus
(rockweed)

An ecologically important seaweed is *Fucus vesiculosus*, which is one of the rockweeds. As its common name suggests, it attaches its holdfast to a strong substrate, such as a jetty, or the banks of a salt marsh. This species of *Fucus* is characterized by two opposite air bladders which may appear at several locations along the blade. These air bladders enable the alga to float in order to photosynthesize more efficiently. Brown in color, the blade also has a distinct midrib. The tips of the alga are swollen when fertile, and these structures release the reproductive cells into the water. Other species here include *Fucus spiralis* in which air bladders are absent, and *Fucus serratus* which has flattened, toothed blades.

Rockweeds may be very abundant in some areas. Consequently, they are microhabitats for small animals such as periwinkle snails. Rockweeds are important in providing food, shelter, and protection for several species. In salt marshes, they often cover large clumps of ribbed mussels.

Leathesia difformis (sea potato)

One of my favorite seaweeds is the strangely shaped brown alga *Leathesia difformis*, commonly called the sea potato. It's one of the easiest to identify, looking puffy and round. It is three to five inches in size, and is usually found washed up on the beach during the warmer, summer months. It disappears in the winter as it enters its microscopic stage.

The sea potato is epiphytic, growing on other seaweeds, particularly the common red alga, Irish moss (*Chondrus crispus*), but I have also seen it growing with eel grass (*Zostera marina*). However, it is typically seen isolated from its host when found on a beach. The round, yellowish-brown lumps are sometimes seen tangled with other plants and shells in the wrack line. Because of its round shape, it does not make a good pressed specimen, and should be preserved in a jar of formalin instead.

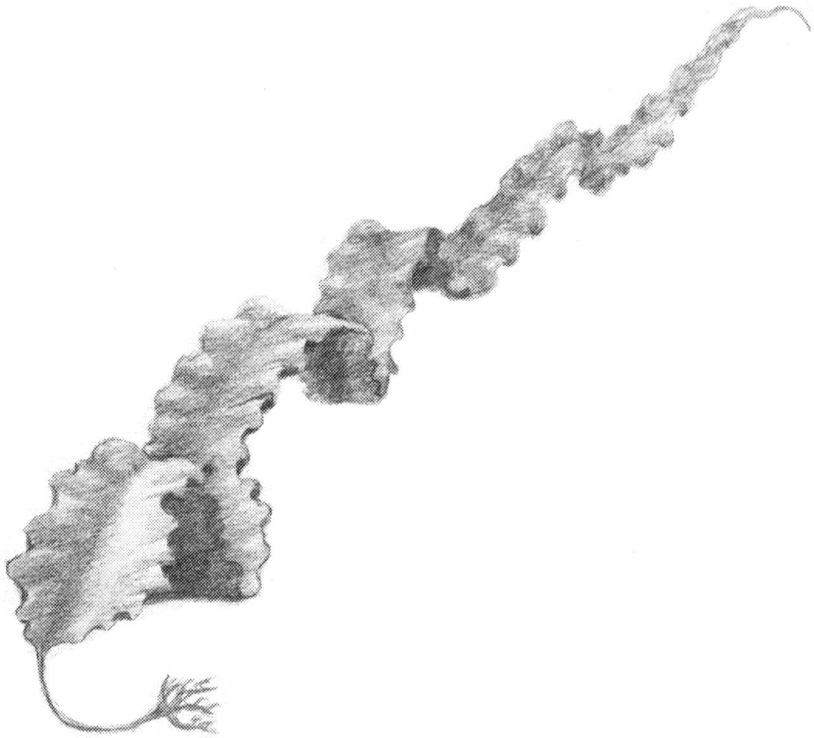

Laminaria saccharina **(kelp)**

One of the largest marine algae species on the east coast is *Laminaria saccharina,* commonly called kelp. The flat blades may be several feet long. Kelp is attached to rocks in deeper cold water by means of a strong, root-like holdfast.

Kelp can be collected year round, and sometimes gets ripped from its habitat by strong currents and storms. Because the blades are so thick, this is a difficult alga to press. One of its Pacific coast relatives is *Macrocystis* which has large air bladders, grows over a hundred feet long, and forms dense forests rich in animal life.

***Scytosiphon lomentaria* (sausage alga)**
Petalonia fascia

Another interesting brown alga is *Scytosiphon lomentaria* (left), the sausage alga. Examine this alga with a hand lens to observe the periodic indentations which gives it the appearance of a group of sausages. Growing up to two feet long, the sausage alga attaches to rocks in the lower intertidal zone.

This thin, hollow alga grows mainly in the winter, and usually disappears in the summer. It often grows near another winter species, *Petalonia fascia* (shown at right), which consists of flat, narrow blades a few inches long. I have collected these two algae species together during the winter. Though *Petalonia* looks like a small piece of kelp, it lacks the large holdfast.

BROWN ALGAE

Sargassum filipendula (Gulfweed)

One brown alga with plant-like characteristics is *Sargassum*, or gulfweed. Most species in this genus are free-floating, but *S. filipendula* is attached. This species is light brown with jagged, narrow "leaves." It also contains many small air bladders throughout the blade that look like tiny fruits.

Sargassum usually grows one to two feet long. Some colonies of this alga contain a large assemblage of animals that are camouflaged in the blades. *Sargassum* may wash up in large concentrations on some Cape beaches, particularly after a storm.

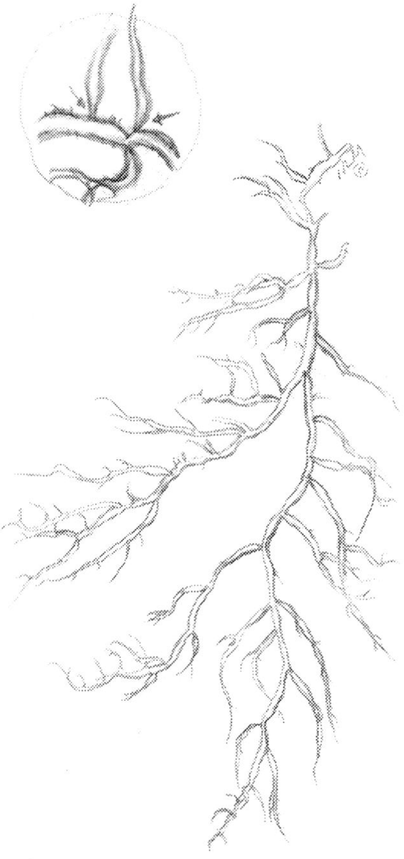

Agardhiella tenera
Gracilaria sp.

Agardhiella tenera, a bushy red alga, is found in quiet waters during the warm months of the year. It can be identified as thickly branched and deep red. Unfortunately, it can be easily confused with other red algae, particularly *Gracilaria sp.* (shown at right). The best way of distinguishing these two common species is to examine the branches closely with a hand lens. Look at the base of several branches. If they taper at the end, you have *Agardhiella.*

The alternate branches of *Agardhiella* are also quite distinct, and I have found this alga in large clumps or mats in bays and estuaries. It's quite attractive underwater with its bright red tinge.

Bonnemaisonia hamifera

You may encounter *Bonnemaisonia*, a small red alga, during the summer even though it grows in the subtidal zone. It breaks apart easily and small clumps will wash up with other algae at high tide. It can be confused with other red algae until you look carefully with a hand lens. Along the branches are small hooks which are plentiful enough to identify this alga.

Whenever I have collected this alga, it is usually entangled with other species. Though the hooks are a clear characteristic, there is another red alga, *Hypnea*, which also has that feature. However, *Hypnea* is a larger alga with the hooks at the tips of the branches.

Ceramium rubrum

There are many small, bushy red algae species, but one that is easily identified is *Ceramium*. I have collected this plant in tide pools, and quiet bays, but it grows primarily in the intertidal zone attached to rocks. Using a hand lens, look for two particular features. The tips fork inward like tiny claws. The bushy branches consist of tiny red bands that encircle the main axis. It is necessary to look for these characteristics because the alga can appear in clumps, or pieces along the beach.

Identifying the genus is easy, but determining the species is more challenging. *Ceramium rubrum* is the most common species in the northeast. Like many red algae species, this alga is more attractive in the water than on the beach. However, because of its soft, fine filaments, it makes an excellent pressed specimen.

Chondrus crispus (Irish moss)

This common red alga, also called Irish moss, is one of the most important commercial algae in this area. An extract from its fronds has been used for food and in various industrial products. *Chondrus* is a deeply red to purple alga with many flat blades up to four inches long. It often grows in large concentrations that wash up into the wrack line after a storm.

Because it is often found along a beach, its exposure to the sun causes the branches to bleach out their pigments giving some of them a whitish to lightly purple appearance before drying out. This species makes an excellent pressed specimen. If collected and pressed fresh, it can retain its color in the herbarium for many years.

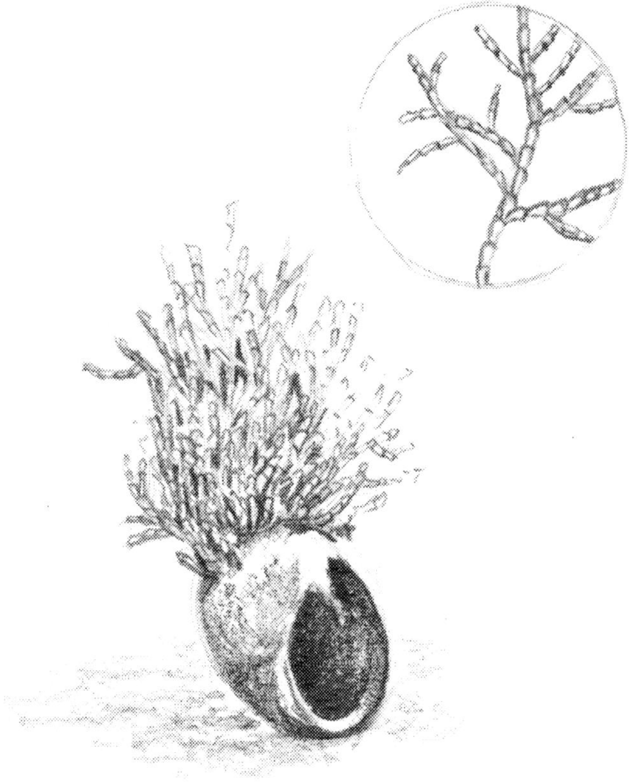

Corallina officinalis (Coral weed)

One common red alga which grows in the lower part of the intertidal zone is *Corallina officinalis,* also called coral weed. Look for a small jointed alga that seems to be encased in a hard, white coating of lime. These algae, like many of their tropical counterparts, have the ability to remove calcium carbonate (lime) from the water. When dead, they are all white. It's this form that you are most likely to find.

Even though they may be covered with lime, the algae are still able to photosynthesize, though probably at a reduced rate. *Corallina* can sometimes be collected attached to small snails or rocks. After a storm, large amounts of this interesting alga can be found entangled with other seaweeds in the wrack line.

Dasya pedicellata (Chenille weed)

One of the prettiest seaweeds is the red alga *Dasya*, which appears as a finely branched and delicate bright red form under water. This species grows when the water is warm, and can be found detached and floating in bays and estuaries.

Upon close examination with a hand lens, you can observe many tiny hairs along the branches. The alga is soft to the touch, and collapses when removed from the water. However, this is one species that should be added to your collection as it is easily pressed, and will retain its color and shape for many years.

Grinnellia americana (Grinnell's pink leaf)

The common name of this red alga describes it perfectly. Looking like a thin pink leaf, *Grinnellia* is a warm water alga a few inches wide. If you remove it from the water, you can see and feel how thin it is. It's because of this characteristic that *Grinnellia* makes an excellent pressed specimen.

Examine this alga closely with a hand lens. Some of them have spots on the blades which are the reproductive spores. *Grinnellia* is found in quiet, warm bays, and is attached to rocks and shells, though detached pieces are often found.

Polysiphonia sp. (Tubed weeds)

A very common and complex red alga is *Polysiphonia*. Its name is quite descriptive. This finely branched alga is composed of many tubes with red bands around a central cell. This can usually be seen with a hand lens, but a look under a microscope will easily confirm this identification.

Polysiphonia is very common, and washes up in bundles with the tide. Like so many other red algae species, it collapses when removed from the water. Known for its complicated life cycle, it is a very diverse alga in which there are over 200 species known throughout the world.

Porphyra umbilicalis **(Nori or laver)**

An economically important red alga is *Porphyra*, known as nori in Asia. This alga looks like a red or purple sea lettuce. The blades are thin, and may be a few inches across. It has been grown and cultivated in Asia for centuries. It can be purchased in health food stores in the United States in a dried form.

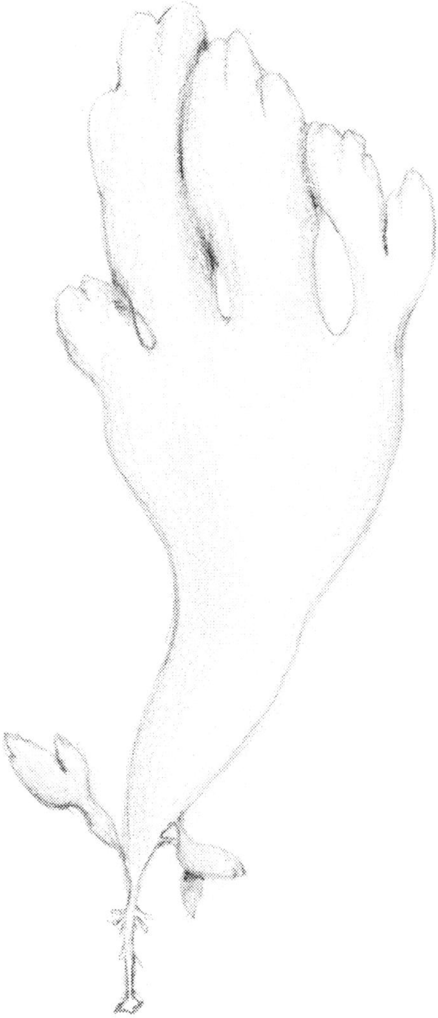

***Palmaria palmata* (dulse)**

This red alga is another economically important species. *Palmaria*, commonly called dulse, has historically been used as a food and a medicine. It is rich in vitamins, iodine, and other minerals, and can be eaten in a raw or dried form. It grows in deeper water attached to rocks and shells, and prefers a rocky shoreline, though it can be found in Cape waters. The blades are reddish-purple, and are shaped like a flattened fan. If collected and pressed in good shape, it can retain its bright color.

FLOWERING PLANTS

Zostera marina **(eel grass)**

Zostera marina, which is also called eel grass, is neither a seaweed nor a grass. However, it is a flowering plant, and therefore has true leaves, stems, roots, flowers, fruits, and seeds. It grows underwater and spreads by an underground stem called a rhizome. You may find it along the beach in large mats. The green leaves are long, and grass-like, but it belongs to the pondweed family. When the plant dies, the leaves turn black, and this is what you will probably find entangled with the seaweed.

Eel grass is very important in the environment. Many species of animals, such as scallops and blue crabs, depend on the thick beds for shelter and food. The plants also stabilize the bottom, and break up wave energy along the shore. With your hand lens you may find several small animals and algae attached to the plants.

Recommended References

Abbott, Isabella A. and E. Yale Dawson. *How To Know The Seaweeds.* Second Edition. Wm. C. Brown Company publishers. Dubuque, Iowa. 1978.

Hillson, C.J. *Seaweeds – A Color-coded, Illustrated Guide to Common Marine Plants Of The East Coast of the United States.* The Pennsylvania State University Press. 1982.

Kingsbury, John and Philip Sze. *Seaweeds of Cape Cod and the Islands.* Bullbrier Press. Jersey Shore, Pennsylvania. 1997.

Petry, Loren C. and Marcia G. Norman. *A Beachcomber's Botany*. The Chatham Conservation Foundation, Inc. Fifth Edition. Chatham, Massachusetts. 1982.

Taylor, William Randolph. *Marine Algae of the Northeastern Coast of North America.* Second Edition. University of Michigan Press, Ann Arbor. 1957.

The Author

Gilbert Newton is a Cape Cod native who has been teaching environmental science at Sandwich High School and Cape Cod Community College for many years. His subjects include botany, coastal ecology, coastal zone management, and environmental technology. He is the past president of the Thornton W. Burgess Society, the Association to Preserve Cape Cod, and the Sandwich Education Association. Gil is also a founder of the Barnstable Land Trust, the past chairman of the Barnstable Conservation Commission, and the director of the Advanced Studies and Leadership Program for the Cape Cod Collaborative and the Massachusetts Maritime Academy.

The Illustrator

While growing up on Cape Cod, Nancy Minnigerode developed a deep love of the natural world and strives to express this love in her art work. Although largely self-taught, Nancy has a degree in Visual Arts from the Cape Cod Community College. Nancy has been a member of Cape Cod Waterworks for over ten years, and has exhibited her work in several art galleries around Cape Cod.

This book was designed and typeset by Nancy Viall Shoemaker of West Barnstable Press, West Barnstable, Massachusetts. The text font is Tiepolo Book, designed by Cynthia Hollandsworth of International Typeface Corp. in 1987 as a "sans serif with serifs alternative" and named after the Italian artist Dominic Tiepolo. The title was set in Ad Lib, designed by Freeman Craw in 1961.

Entire book, including cover, printed on recycled paper